Rafaela Mazzutti

Rainfall variability and trends in the Rio Doce/MG basin

Rafaela Mazzutti

Rainfall variability and trends in the Rio Doce/MG basin

An analysis of the rainfall regime and its future trends in the river basin

ScienciaScripts

Imprint

Any brand names and product names mentioned in this book are subject to trademark, brand or patent protection and are trademarks or registered trademarks of their respective holders. The use of brand names, product names, common names, trade names, product descriptions etc. even without a particular marking in this work is in no way to be construed to mean that such names may be regarded as unrestricted in respect of trademark and brand protection legislation and could thus be used by anyone.

Cover image: www.ingimage.com

This book is a translation from the original published under ISBN 978-613-9-62607-6.

Publisher:
Sciencia Scripts
is a trademark of
Dodo Books Indian Ocean Ltd. and OmniScriptum S.R.L publishing group

120 High Road, East Finchley, London, N2 9ED, United Kingdom
Str. Armeneasca 28/1, office 1, Chisinau MD-2012, Republic of Moldova, Europe
Printed at: see last page
ISBN: 978-620-7-74606-4

SUMMARY

THANKS

I would first like to thank my parents for their love, encouragement and unconditional support in all my choices. Mom, your care and dedication was essential and, at times, was my strength and hope to keep going. Dad, your support has meant security and the certainty that I am not alone on this journey.

To my advisor and friend Vanderlei, who was always there to help with questions and doubts. I would like to thank him for his commitment to this academic work.

To my sister, Cris, who is my partner, my friend and who, in a very special way, has always been present in my life and in my choices. It's a good thing that you don't choose your family.

To my friend Murilo, who supported me at all times and gave me the strength to continue on my path.

My daughter Betina, who is the great love of my life.

SUMMARY

Rain is one of the main elements necessary for the existence of life on planet Earth. In order to understand its dynamics, we need studies and diagnoses that analyze the variability of its occurrence and its probable future trends in relation to a time frame, so that they can be used as a tool to determine parameters of climatological variability. With the data studied, rainfall trends can be used to guide human activities such as agriculture and improve land use in a given area. The aim of studying rainfall variability in the Doce River basin is therefore to try to determine a climatological rainfall pattern for the period from 1941 to 2002, based on the spatio-temporal distribution of 46 meteorological stations installed throughout the basin. As a result, it was possible to characterize the temporal rainfall regime, including verification of the intense variability, as well as obtaining a spatial visualization of the distribution of rainfall in the basin.

Keywords: Trends, rainfall variability, Doce river basin

1. INTRODUCTION

The hydrographic basin can be characterized as a natural geographic compartmentalization, which is responsible for bringing together various natural elements that make up the landscape. Among these, the climate stands out for its environmental characteristics which are intrinsically related to society, insofar as it interacts with human activities.

Ayoade (1996) considers that the interface between climate and society is related to vulnerability and the prevention of climate impacts. A society is more vulnerable when: the more its economic activity depends on climate-sensitive factors of production; the less it depends on certain essential climatic variables, such as rainfall and temperature; the lower its capacity to reserve materials to help the homeless, victims of natural disasters; the less prepared it is to deal with adverse climatic impacts.

Tavares (2004), referring to the relationship between man and climate, within the context of climate change, states that living beings, morphogenetic processes, river regimes and the activities carried out by man are linked to the current atmospheric situations, understood as essential to shaping the climate, and that the intensity of rainfall episodes leads to greater damage, such as soil erosion or the need to open dam floodgates.

As for climate, Nimer (1989) states that it depends on static factors (the physical conditions of the planet) and dynamic factors (the dynamics of the atmosphere) which define its characteristics. Minas Gerais stands out for its great diversity of climates, as it is a tropical climate transition region. The dynamics of this state's climate originate from global circulation mechanisms, such as tropical atmospheric circulation cells and frontal systems (dynamic factors) and their interaction with the tropical continental climate and the rugged regional topography (static factors).

In this study, the hydrographic basin in question is the Doce River Basin. According to Lage et. al (2005), the Doce is an interstate river, 875 km long, whose source is located in the Serra da Mantiqueira in the municipality of Ressaquinha, Minas Gerais, at an altitude of 1,200 meters. The region has a tropical high-altitude climate with three subtypes: relatively cool summers in the high altitudes, mild summers in the middle altitudes and hot summers in the lower altitudes. Its main sources of water are the rivers: Xopotó, Piranga and Carmo.

The river gets its name from the meeting of the Carmo and Piranga rivers, below the town of Ponte Nova, Minas Gerais, and its mouth is in the municipality of Regência, Espirito Santo. The Doce River basin has a drainage area of 83,400 km2, 86% of which is located in Minas Gerais and 14% in the state of Espirito Santo.

The occupation of this region was due, in the mid-18th century, to the discovery of gold and diamonds in its western part, in the municipalities of Peçanha and Serro, and, from the 19th century onwards, to the search for areas for coffee plantations. Recently, the regional economy has been based on mining, agriculture and industry.

In economic terms, the Doce River basin contributes to the mineral production of mica, colored stones, limestone and ferrous minerals. Agriculture includes the cultivation of corn, beans, coffee, manioc, sugar cane, bananas and rice. Prominent industrial activities in the region are: production of non-metallic minerals; food and beverage production; cellulose and steelmaking. Tourism is also prominent.

The Doce River basin, due to its urban and industrial development since the 1970s and the consequent increase in demand for electricity, has the largest number of small and medium-sized hydroelectric dams (SHPs) in Minas Gerais.

Rainfall is very important for economic activities and environmental balance in river basins. Due to the variability of its occurrence, diagnostics that analyze rainfall distribution over a period of time are essential, so that they can be used as a tool to determine parameters of climatological variability. The study of rainfall is essential for a better understanding of the hydrological cycle. Measuring rainfall is of great importance as it can be used to properly manage a river basin, thus helping to manage the quality of the environment and the dynamic process of various socio-economic activities.

Thus, the aim of studying rainfall trends and variability in the Rio Doce basin is related to the need to identify a possible climatological pattern in the period from 1941 to 2002, according to data from 46 rainfall stations installed throughout the basin. In this study, rainfall data was selected from records held by the National Water Agency (ANA). The data is downloaded from the *Hidroweb* link and then the *HIDRO 1.2 software is* used to tabulate and interpret the information.

Initially, there were a total of 102 rainfall stations in the area covered by the Doce River basin. With these, a *Microsoft Office Excel* spreadsheet was assembled, breaking down the time sequence of the data (historical series). Of these, 30 had no significant data because they had already been deactivated. After filtering, 46 stations were chosen that had consistent records from 1941 to 2002.

Graphs were built for each station from the selected stations. These contain information on annual rainfall totals measured in millimeters of rain. Special attention was paid to checking for a possible increase in annual rainfall totals, with 18 stations in the basin showing this behavior. On the other hand, 18 other stations showed a downward trend in annual rainfall totals and 10 showed stability.

Another important issue that was taken into account was the deviations of the absolute annual rainfall values for each station from the average for the entire period of the respective historical series. The deviation graphs showed the stations with the greatest variability over the period analyzed.

In short, the aim of this work is to gain a quantitative understanding of the rainfall differences in the Doce River drainage basin over 61 years (1941-2002) of data collected. Rainfall variability becomes important when associated with land use and occupation studies, as well as to understand possible changes in the basin's climate pattern.

2. BIBLIOGRAPHIC REVIEW

2.1 . Rainfall Time Series

2.1.1. Time Series Concept

A time series is any set of information ordered in time (Morettin and Toloi, 2006). For Silva *et al.* (2007), it is a grouping of discrete observations made at equidistant times and which show serial dependence between them. The study of time series has great applicability in climatology because it allows the prediction of future values for various climatic elements. Detecting whether there is a progressive increase or decrease in temperature or rainfall on a mesoclimatic scale is fundamental for identifying externalities produced by changes in land use or global climate change on the hydropluvial regimes of river basins (FERREIRA, 2012).

Rainfall and the availability of water resources affect a variety of human activities, including fishing, navigation, public water supply, agriculture and hydroelectric power generation. It is therefore essential to research the variability of rainfall series in order to better plan land use and guide human activities such as agriculture. Rainfall shows significant inter-annual fluctuations; for this reason, criteria and procedures should be proposed to identify regular years, rainy years and dry years in historical series.

2.1.2 Variability in rainfall series in river basins

Rainfall is one of the main elements necessary for the existence of life on planet Earth. Due to the variability of its occurrence, diagnostics that analyze the rainfall distribution of a river basin in relation to a time frame become essential, so that they can be used as a tool to determine parameters of climatological variability.

The aim of studying the rainfall variability of the Rio Doce basin is therefore related to the need to determine a climatological pattern over a relatively long period, according to the spatial distribution of the rainfall stations installed throughout the basin. This makes it possible to visualize the rainfall regime, with its clear rainfall variability.

The hydrographic basin is often used as a geographic reference for the adoption of planning practices or the management and use of natural resources. Given the great importance of water as a route for transportation, for the generation of electricity, as a source of urban and industrial supply and as a route for the dilution of domestic and industrial effluents, the hydrographic basin has become a basic unit for environmental planning and management (ROSS E DEL PRETTE, 1998).

Therefore, knowledge of the predominant rainfall patterns at different scales and their variability becomes even more important in water resources planning, hydrological studies, urban planning and agricultural planning, among others, in river basins.

With regard to rainfall variability, several related factors are taken into account, from anthropogenic actions to large and mesoscale atmospheric processes such as El Nino, La Nina, convergence zones and air masses. Other factors that can be related to rainfall variability are physiographic and morphological characteristics, vegetation cover and land use.

Taking these factors into account, calculations such as arithmetic mean, standard deviation and coefficient of variation help to detect trends, variations and spatial and temporal similarities or differences. For them to be reliable, it is necessary to use a wide variety of stations with complete long-term data.

Generally speaking, despite the fact that some local differences lead to variations in rainfall, the stations used in this study are under the influence of the same atmospheric processes. From this perspective, the trends found in the distribution and variability of rainfall can be used to avoid various urban and rural problems such as flooding and erosion, and can contribute to urban planning and land use initiatives.

2.2 Characterization of the study area

2.2.1 Occupation process

According to Lage et. al (2005), the Doce is an interstate river, 875 km long, whose source is located in the Serra da Mantiqueira in the municipality of Ressaquinha, Minas Gerais, at an altitude of 1,200 meters.

The river gets its name from the meeting of the Carmo and Piranga rivers, below the town of Ponte Nova, Minas Gerais, and its mouth is in the municipality of Regência, Espirito Santo. It has a drainage area of 83,400 km2, 86% of which is located in Minas Gerais and 14% in the state of Espirito Santo (FIG. 1).

Figure 1 Location map of the Doce River basin Source: PIRH - Doce River Basin, 2010.

The main geographical boundaries of the basin are: to the north, the Negra and Aimorés mountains; to the west, the Espinhaço mountain range; to the southwest and south, the Mantiqueira mountain range; to the southeast, the Caparaó mountain range; to the east, the Atlantic Ocean.

According to Strauch (1955), human occupation of eastern Minas Gerais and the Rio Doce valley took place in two different directions and at different times. From the plateau towards the coast, the mining cycle took place. From the coast, towards the interior, the occupation of agricultural land. Quoting Prado Jùnior (1953), the author refers to this region of Minas Gerais as "a settlement that was much more intense there and, above all, was organized on a more solid social basis. It was therefore possible to renew and reconstitute, partially at least, the lost mining sector with other elements of vitality: livestock and agriculture."

Ribeirâo do Carmo and Vila Rica, today Mariana and Ouro Preto, were the gateway to the occupation of the basin. However, in order to avoid the "gold trails", the Portuguese Crown banned navigation on the Doce River. In addition to the prohibition, the closed forest, malaria and the Botocudo Indians, known for their hostility, made the region one of the last to be occupied in Minas Gerais. Until the beginning of the 20th century, the Doce River valley remained largely covered by the Atlantic Forest complex (CBH-Doce, 2001).

The region was only effectively occupied when the Vitória-Minas Railroad (EFVM) was built. Started in 1903 in Vitória, in 1910 it reached the then small trading post of Porto de Figueiras, today Governador Valadares.

In the 1930s, the EFVM reached Itabira, in the Piracicaba river basin, from whose mines iron ore would be extracted for export via the Port of Vitória. In 1937, the first steel mill was set up on the banks of the Piracicaba River, Companhia Siderùrgica Belgo Mineira. In 1942, Companhia Vale do Rio Doce was created in Itabira.

The 1930s also saw the introduction of colony grass in the Porto Figueiras region, which made it possible to expand livestock farming. The introduction of pastures and the strong demand for coal for the steel mills and wood - with post-war Europe, the United States and Japan as major consumers - led to a major process of deforestation in the basin.

At the beginning of the 1950s, the Rio-Bahia highway was opened, passing through Governador Valadares, making it a migratory corridor for people from the Northeast region. In addition, as the land after the clearing of the forest was not suitable for much more than livestock farming, the city of Governador Valadares experienced a demographic swelling, reaching a growth rate of 13.3% in the decade (IBGE).

In 1953, also on the banks of the Piracicaba, the Companhia de Aços Especiais Itabira - ACESITA was inaugurated. Ten years later, a few kilometers downstream, the Usina Intendente Câmara - USIMINAS - began operating. The installation of the steel mills led to the emergence of the Steel Valley Urban Agglomeration, involving the cities of Ipatinga, Coronel Fabriciano and Timóteo. In the 1940s, eucalyptus was introduced to the region as a way of relieving the pressure on natural forests.

Following the course of their history, the farms set up at the headwaters of the Doce basin, still in the gold rush, continue to survive on what they produce. The massive eucalyptus forests in the region made it possible in 1975 to set up Companhia Nipo-Brasileira - CENIBRA, a pulp producer located on the banks of the Doce river, downstream from the mouth of the Piracicaba river, in the municipality of Belo Oriente.

The socio-economic consequences are diverse and complex. In rural areas, for example, there are dozens of small conflicts between farmers who, in an attempt to solve their own problems, end up interfering with the entire watercourse downstream.

As a result, the middle Doce River region alone (Tumiritinga to Aimorés) - which is in an advanced state of desertification - lost around 40% of its population between the 1970s and 1980s (IBGE). In Minas Gerais, the Doce river basin is characterized as the region that lost the most population: 615,259 inhabitants between 1970 and 1991 (UFMG, IBGE).

Covering two hundred and thirty municipalities in the eastern region of Minas Gerais and part of Espírito Santo, the Doce River basin's economy is based on a mosaic of activities: large mining projects; eucalyptus forestry; large steel mills; hydroelectric power generation; the exploitation of precious and semi-precious stones; beef and dairy cattle farming; pig farming; sugar cane; coffee growing; as well as subsistence farming. Therefore, this basin not only plays an important role in Minas Gerais' economy, but also in Brazil's.

Despite the various advances in environmental issues in the region, rapid economic growth and improved infrastructure have been accompanied by significant environmental impacts, such as the accelerated and unsustainable increase in the demand for natural resources, high levels of atmospheric and water pollution and soil loss, which contributes to the silting up of rivers.

Guerra and Barbosa (1996) state that the short-sighted regional economic development model implemented in the Doce basin, especially since the 1970s, has led to a high degree of environmental degradation, negatively influencing the complex interactions that exist in the socio-environmental dynamic. Among the consequences of this degradation are the urban floods that occur in several municipalities as a result of the interaction between the heavy rains in December and January and the irregular use and occupation of the land by the population.

Most of these municipalities occupy in a disorganized way the space that rightfully and de facto belongs to the rivers, i.e. their floodplains. In this way, the populations of these cities live with the expectation, during the rainy season, of new floods, which cause enormous material damage in all socio-economic sectors.

2.2.2 Weather

Starting the discussion with a broader approach, it is understood that the Doce River basin is under the influence of the regional atmospheric dynamics that act on the state of Minas Gerais and, on a larger scale, on Brazil. Nimer (1989) states that the southeast of Brazil, due to its latitudinal position, is characterized by being a transitional region between the warm climates of low latitudes and the temperate mesothermal climates of mid-latitudes. The macro-climate of Minas Gerais is characterized by a seasonality that is responsible for two distinct and well-defined seasons, a hot, humid summer and a mild, dry winter, as well as two transitional seasons, autumn and spring.

Among the climatic influences to which the whole country is subjected is the South Atlantic Convergence Zone (SACZ). Silva Dias and Marengo (2002, p.71) state that:

> The South Atlantic Convergence Zone (SACZ) is conventionally defined as a band of persistent cloudiness oriented in a northwest-southeast direction, associated with a convergent flow in the lower troposphere, extending from the south of the Amazon to the South-Central Atlantic for a few thousand km, well characterized in the summer months. The SACZ has gained prominence in recent years as an efficient meteorological system that produces heavy rainfall, and deserves special attention due to the complexity of its structure and the controversy surrounding the mechanisms of its formation, maintenance and destruction.

According to Cupolillo (2008, p. 29) the SACZ is made up of a combination of atmospheric mechanisms acting on the South American continent. These mechanisms are: the Bolivian High, continental tropical convection originating in the Amazon and frontal systems coming from the southern part of the continent. In this way, a NW-SE band of cloudiness forms over South America. At the surface, the humid air is transported from the Amazon to the southeast of Brazil, reaching the region of the Doce River basin. During the rainy season, the SACZ is often stationed over the 19th and 20th parallels of south latitude, corresponding to the location of the Doce River basin, causing disasters in many municipalities in the basin, such as floods and collapsed barriers on highways and in urban areas.

With regard to local climatic aspects, the basin, like the state of Minas Gerais (NIMER, 1989), is influenced by the Atlantic Tropical (ATM), Atlantic Polar (APM) and Continental Equatorial (CE) air masses, characterized by westerly currents. The MEC acts in spring and summer, causing high average annual temperatures, both minimum and maximum. The region is influenced by the sea, causing higher temperatures in Baixo Guandu, in Espírito Santo, and in Aimorés, Governador Valadares and the Vale do Aço region (Timóteo, Coronel Fabriciano and Ipatinga), in Minas Gerais. However, for the whole basin, the temperature is milder in the upper Doce River. The rainfall regime has two well-defined periods: rainy, from October to March, and dry, from April to September. The total accumulated rainfall of twelve hundred millimeters is distributed over the period from October to March, and is concentrated in December, January, February and March.

According to Koppen's climate classification, the following types of climate can be found in the Rio Doce basin area: Am (hot and humid), Aw (hot with summer rains), Cwa (high altitude tropical with summer rains: hot summers), Cwb (high altitude tropical with summer rains: cool summers), Cfa (subtropical with distributed rains and hot summers) and Cfb (subtropical with distributed rains and cool summers). Figure 2 shows the distribution of climatic types in the basin.

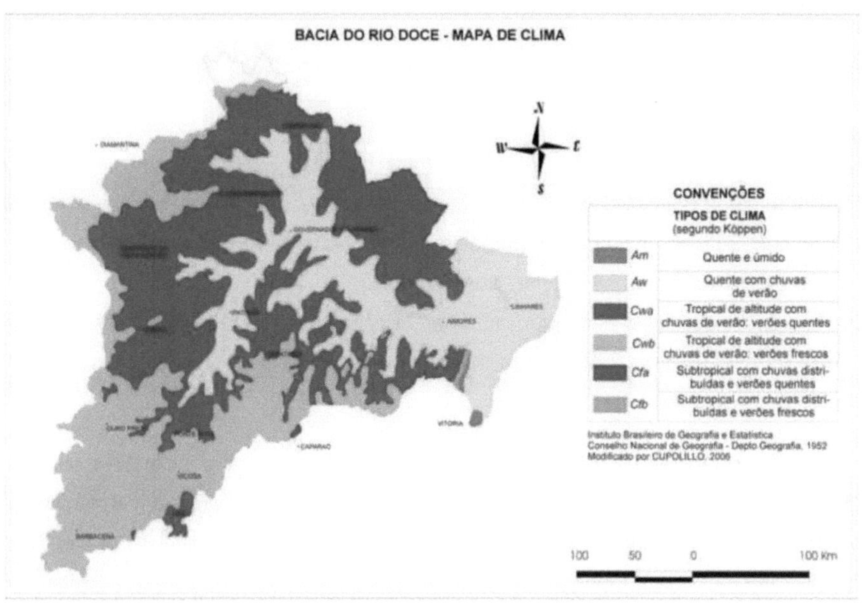

Figure 2 Map of the W. Koppen climate classification for the Doce River basin. Source: CUPOLILLO,2008

With regard to the spatial distribution of rainfall, the highest rainfall values are located in the western region of the basin and the lowest rainfall values in the eastern region of the basin. The region characterized as the least rainy in the basin is represented by a group of terrains with modest altitudes or even at sea level. This compartment includes the coastal plain domains and the tablelands, which mark the transition from this geomorphological unit to the collinear domains. (CUPOLILLO, 2008) In this topographic compartment, the additional characteristic, apart from the low altitudes, is the presence of a surface of low or zero roughness, which influences the path of the air masses that circulate in this domain.

On the other hand, a wider compartment of the Doce River basin that extends from the vicinity of its mouth to the base of the mountains that mark its eastern boundary are the domains of the mares de morros. Here, the altitudes rise slightly towards the west. On the western border, there are a number of continentally influenced mountain ranges, such as the Mantiqueira and Espinhaço ranges, which are important frontiers for the action of air masses generated in the Atlantic Ocean. In general, these aspects related to the altitudes of the basin should influence the behavior of the air masses in the regional context, explaining the rainfall dynamics in the region.

2.2.3 Physiographic compartmentalization

Three major compartments of the geomorphological aspects should be highlighted. These are the major topographic transition marks in the Doce River basin, which allow for the identification of easily delimited spatial units.

The first of these is represented by a group of terrains at modest altitudes or even at sea level. This compartment includes the domains of the coastal plain and the tablelands that mark the transition from the former to the collinear domains. In this topographic compartment, the additional characteristic, apart from the low altitudes, is the presence of a surface of low to zero roughness, which influences the path of the air masses that circulate through this domain.

The second compartment is a little more complex as it includes structures with obvious differences from the larger spatial domain considered here. This compartment can be considered the largest in the Doce River basin and extends from the vicinity of its mouth to the base of the mountains that mark its eastern boundary. It is a vast area dominated by "half-orange hills", which typify the mares de morros domain. It is a domain whose altitudes increase discreetly towards the west and is clearly rugged. As mentioned above, this domain is home to mountain ranges, some of which are more specific in size, others of regional scope, such as the Serrano do Caparaó domain. In general, from the point of view of climatological dynamics, regardless of the morphogenetic context that characterizes it, these aspects should influence the behaviour of air masses in the regional context.

Finally, it should be noted that the third compartment to be highlighted is represented by mountain ranges with continental influence, such as the Mantiqueira and Espinhaço ranges, which are important frontiers for the action of some air masses generated in the

Atlantic Ocean.

2.2.4 Land use

According to research carried out by the Minas Gerais Technology Center Foundation (CETEC), 95% of the land in the basin is pasture and woodland, demonstrating the predominance of livestock farming. The most widespread species used to form pastures are fat grass (Melinis minutiflora) in areas located above 800 m above sea level and colony grass (Panicum maximum) below this altitude. Planted forests, consisting mainly of species of the Eucalyptus genus, are significant in the middle Rio Doce. Almost all the reforestation belongs to the Acesita and Belgo Mineira steel companies or to Cenibra, a pulp producer. There are fewer fields and cultivated areas. Due to the characteristics of the soils in the Doce River basin and inadequate management, erosion has become one of the biggest environmental problems in the region.

In the countryside there are vast areas in an advanced state of desertification, eutrophied lakes, unprotected springs and erosion processes. More than 90% of the original vegetation cover is extinct. Of the remainder, less than 1% is in a primary stage (Mittermeier et alli, 1982; Fonseca, 1985).

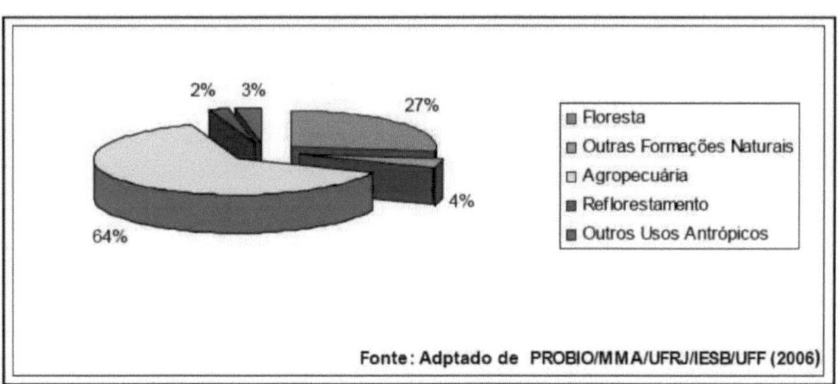

Figure 3 Land use and occupation by typology in the Doce River basin.

Source: PIRH Doce River Basin, 2010

According to the data mapped and represented in this figure, most of the Doce River basin is anthropized, especially by agricultural use. Forests, which used to cover around 90% of the basin, are now found in less than 1/3 of the total area. However, the quality of these forest fragments is not really known, so the mere presence of the *Forest* class does not mean that it presents sufficient phytostructural conditions to ecologically maintain the local flora and fauna (SOS MATA ATLÂNTICA / INPE, 2001).

3. METHODOLOGICAL ROADMAP

3.1 Data collection and basis

The rainfall data used in this study was selected from 46 points in the records of the National Water Agency (ANA). The data covers a total period of 61 years (1941 to 2002) and is available on the *Hidroweb* tool on the ANA website. After retrieving them from the website, they were tabulated and processed using the *HIDRO 1.2 software.*

Initially, there were a total of 102 climatological stations in the area covered by the Doce River Basin. An Excel spreadsheet was created with these stations, breaking down their numbers and years of data. Of these, 30 had no significant data due to the fact that they had already been deactivated and had a reduced number of years. After filtering, 46 stations with records from 1941 to 2002 were chosen (table 1).

Using *ArcGIS software*, a map was created showing the exact location of the stations within the basin boundaries. With this information, it was possible to assess the spatial distribution of rainfall in the region and understand its dynamics (FIG. 4).

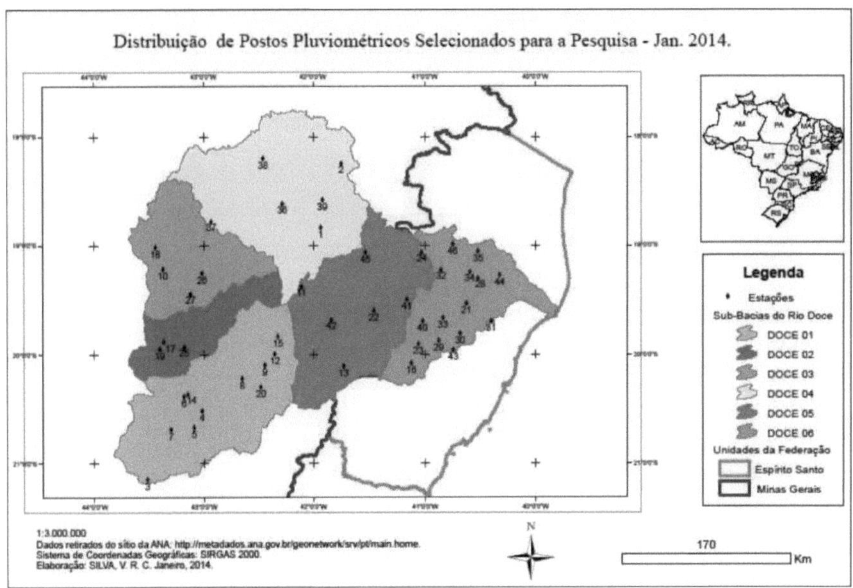

Figure 4 Distribution of rainfall stations selected for the research.

Source: From the author

Numbered Posts	Codes	Names	Coordinates
1	1840000	White Eagle	18°59'8"S 40°44'46"W
2	1841001	Vila Matias	18°34'29"S and 41°55'4"O
3	1841003	Campaign	18°14'19"S 41°44'55"W
4	1841004	Gov. Valadares	18°50'0"S and 41°56'0"O
5	1842005	Coroaci	18°36'43"S 42°16'43"W
6	1842007	Guanhaes	18°46'20"S and 42°55'52"O
7	1842008	St. Maria do Suaçui	18°12'4"S 42°27'19"W
8	1940000	Itarana	19°52'28"S and 40°52'28"O
9	1940001	Sao Joao de Petrópolis	19°48'19"S 40°40'44"W
10	1940005	Horse	19°41'32"S and 40°23'53"O
11	1940006	Colatina	19°31'51"S and 40°37'23"O
12	1940009	Pancas	19°13'13"S and 40°51'12"O
13	1940012	Itaimbé	19°39'49"S and 40°50'7"O
14	1940013	New Brazil	19°14'15"S and 40°35'29"O
15	1940016	Barra de Sao Gabriel	19°3'28"S and 40°30'59"O
16	1940020	Boiler	19°57'18"S and 40°44'30"O
17	1940023	Bananal River	19°16'27"S and 40°19'15"O
18	1940025	Sawmill	19°17'43"S and 40°31'3"O
19	1941005	Barra do Cuieté	19°3'42"S and 41°31'58"O
20	1941006	Assarai	19°35'41"S and 41°27'29"O
21	1941008	Orange of the Earth	19°54'4"S and 41°3'29"O
22	1941009	Ibituba	19°41'28"S and 41°1'12"O
23	1941010	Sao Sebastiao da Encruzilhada	19°29'33"S and 41°9'42"O
24	1941011	St. Antônio do Manhuaçu	19°40'42"S and 41°50'10"O
25	1941012	Upper Rio Novo	19°3'33"S and 41°1'39"O
26	1942002	Bom Jesus do Galho	19°50'1"S and 42°19'4"O
27	1942006	Old Red	19°59'56"S and 42°20'51"O
28	1942008	Dom Cavati	19°22'25"S and 42°6'18"O
29	1943001	Picacicaba River	19°55'22"S and 43°10'40"O
30	1943002	Conceiçao do Mato Dentro	19°1'0"S and 43°26'39"O
31	1943003	Irons	19°15'1"S and 43°0'52"O
32	1943007	St. Barbara	19°56'43"S and 43°24'4"O
33	1943008	Sta. Maria do Itabira	19°26'25"S and 43°7'7"O
34	1943025	Morro do Pilar	19°13'3"S and 43°22'27"O
35	1943027	Peti Plant	19°52'51"S and 43°22'3"O
36	2041008	Dores do Manhumirim	20°6'29"S and 41°43'42"O
37	2041023	Afonso Claudio	20°4'43"S and 41°7'17"O
38	2042008	Raul Soares	20°6'13"S and 42°26'24"O
39	2042010	Open Field	20°17'56"S and 42°28'41"O
40	2042011	Rio Casca	20°13'34"S and 42°39'0"O
41	2043009	Acaiaca	20°21'45"S and 43°8'38"O

42	2043010	Piranga	20°41'26"S and 43°17'58"O
43	2043011	Paraiso Farm	20°23'24"S and 43°10'49"O
44	2043014	Porto Firme	20°40'13"S and 43°5'17"O
45	2043025	Usina da Brecha	20°31'0"S and 43°1'0"O
46	2143003	Desterro do Melo	21°8'57"S and 43°31'12"O

TABLE 1 Numbered stations, codes, station names and coordinates

3.2 . Preparation of graphs and analysis of rainfall trends and variability

Graphs were built for each station from the selected stations. These show the annual rainfall totals measured in millimeters of rain for each year. With this result, it was possible to visualize and divide the regions that have similar rainfall trends. Among them, three groups were found, with distinct behaviors:

- Increasing trend in annual rainfall totals

- Downward trend in annual rainfall totals

- Trend towards stable annual rainfall totals

In this first group, there are 18 stations out of the 46 analyzed, which are characterized by an increase in rainfall. With the same amount, 18 other stations are characterized by a decrease in the amount of rainfall. As for the stations with a tendency towards stable annual rainfall totals, there are 10 stations with this behavior.

As far as variability is concerned, graphs were made showing the relationship between the amount of rainfall in each year in relation to the average rainfall for all the years in the period for each weather station. They therefore indicate whether there were significant negative or positive anomalies. The average standard deviation was also calculated, which indicates the variability of the rainfall totals in relation to the total average for each station. Values closer to zero indicate that there was no variability, i.e. that in the season in question, rainfall amounts were close to the average. Higher values indicate much higher or much lower rainfall than the average. Lower values indicate values close to the average. The average added to and subtracted from the standard deviation was calculated and the number of years below and above this number was counted, characterizing a greater or lesser number of rainy years.

4. RESULTS

4.1 Rainfall trends using linear regression

In this section, the analysis of the stations with the aim of finding a trend or rainfall behavior will be presented from the perspective of linear regression. They were divided into 3 groups.

4.1.1 Group 1 - Trend of increasing annual rainfall totals

This can be considered the most representative group if we take into account that a total of 18 stations showed this behavior. For the sake of demonstration, six stations that show a trend of increasing annual rainfall totals were selected and analyzed below. They are representative of this behavior.

Station 1940006 - Colatina - Fire Department

Located in the state of Minas Gerais, in the northwest region of the DOCE 05 sub-basin, the station has data collected from 1972 to 2002. Its maximum recorded rainfall was 1535.5mm in 1979 and its minimum was 600.2mm in 1986. According to the linear regression, the annual totals show a slight upward trend, varying on average from 900 to 1100mm.

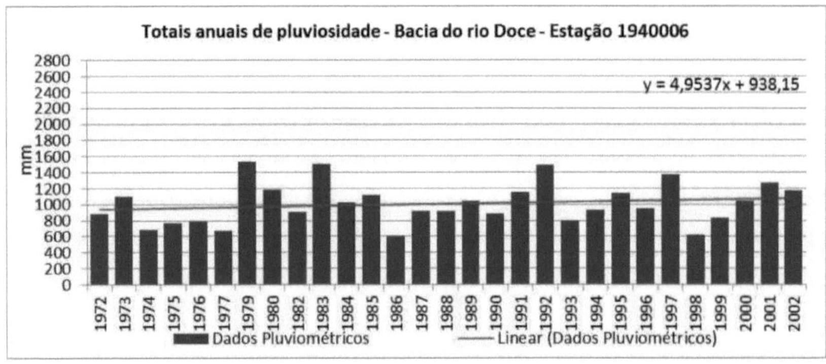

Station 1941005 - Barra do Cuieté – Downstream

Located in the state of Minas Gerais in the eastern region of the DOCE 02 sub-basin, the station has data collected from 1962 to 2002. The maximum recorded was 1665.2 mm in

1991 and the minimum was 211.8 mm in 1963. There was no precipitation from April to October, which characterizes the longest period of drought at the station.

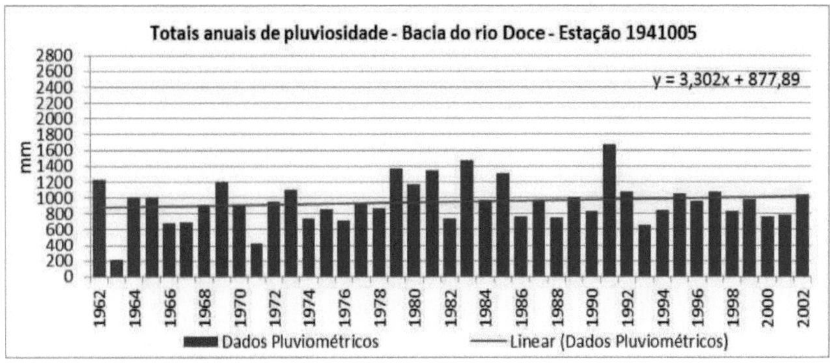

Figure 6 Annual rainfall totals at station 1940006

Station 1941006 - Assarai - Upstream

Located in the state of Minas Gerais in the southwestern region of the DOCE 01 sub-basin, the station has the highest average annual rainfall in 1984, which is 1581.6mm, and the lowest in 1971, 327.3mm, with 5 months of drought. The linear ranges from 900 to 1100mm.

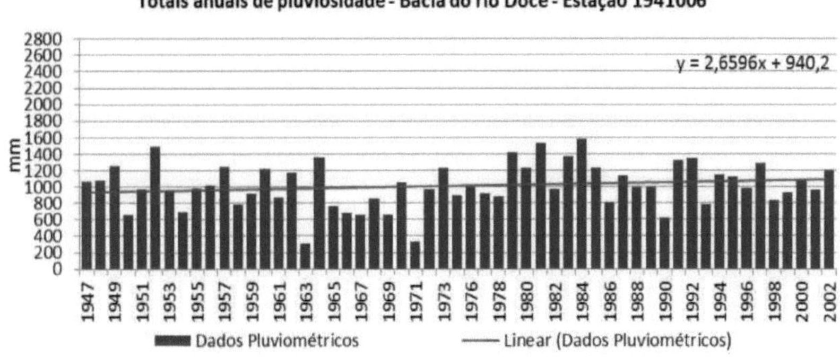

Figure 7 Annual rainfall totals for the 1941006 season

Station 1942002 - Bom Jesus do Galho

Located in the state of Minas Gerais in the western region of the DOCE 02 sub-basin, the station has data collected from 1942 to 2002. This is exactly the year in which the highest annual average was collected, 1660.2mm. The lowest annual total was recorded in 1963 with 354.2mm. The linear ranges from 900 to 1100mm on average.

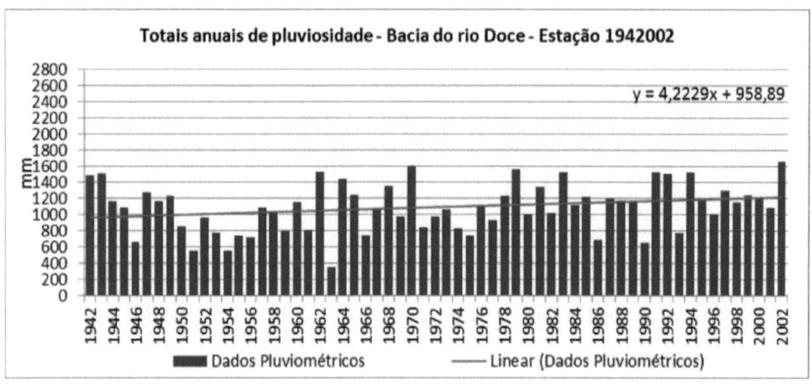

Figure 8 Annual rainfall totals for the 1942002 season

Station 1943001 - Piracicaba River

Located in the state of Espirito Santo in the eastern region of the DOCE 06 sub-basin, the station has an average linear regression between 1300 and 1450mm. The maximum annual rainfall occurred in 1983 at 2027.3mm and the minimum was again recorded in 1963 with 396mm and three months of complete drought.

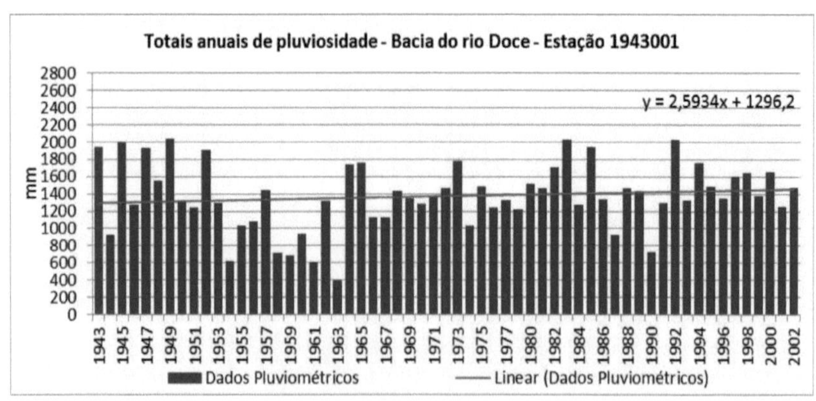

Figure 9 Annual rainfall totals at station 1943001

Station 2043010 - Piranga

Located in the state of Minas Gerais in the eastern region of the DOCE 05 sub-basin, the station has data from 1954 to 2002. In 1983, 2018.3mm was the highest rainfall recorded. The lowest, again in 1963, was 293.6mm, with five months of drought, making it the driest year on record for the entire Doce River basin. The upward trend can be seen in the linear regression ranging from 1000 to 1600mm.

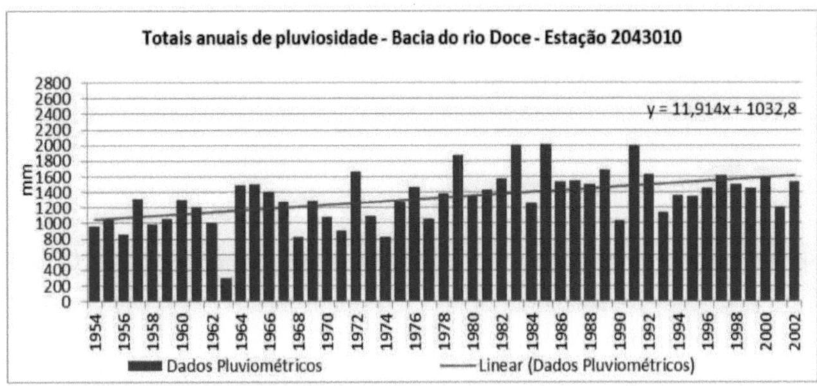

Figure 10 Annual rainfall totals at station 2043010

4.1.2 Group 2 - Stability trend in annual rainfall totals

Of the stations surveyed, six show a trend towards stable rainfall. To illustrate this, another six stations were chosen, demonstrating the stability of rainfall over the years in some regions of the basin.

Station 1841001 - Vila Matias - Upstream

Located in the state of Minas Gerais, in the northern region of the DOCE 04 sub-basin, it has data from 1952 to 2002 showing a linear trend of rainfall stability. The lowest annual value was 319.7mm in 1963 and the highest was 1845mm in 1983.

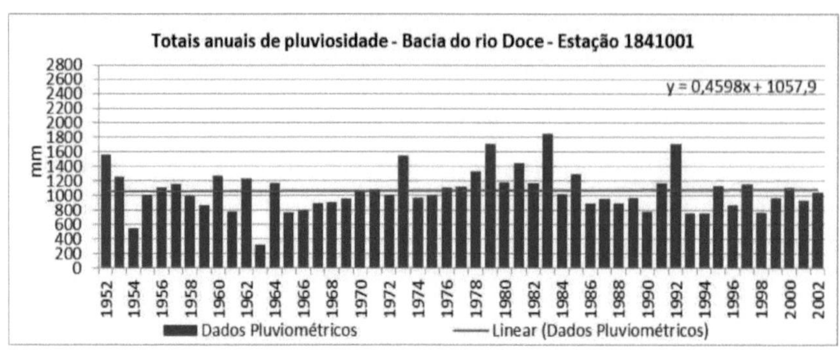

Figure 11 Annual rainfall totals at station 1841001

Station 1841003 - Campanàrio

Located in the state of Minas Gerais, in the southwestern region of the DOCE 01 sub-basin, it has data from 1964 to 2002. Rainfall varies from 249.2 to 1689.5 mm per year. The linear average is close to 900mm.

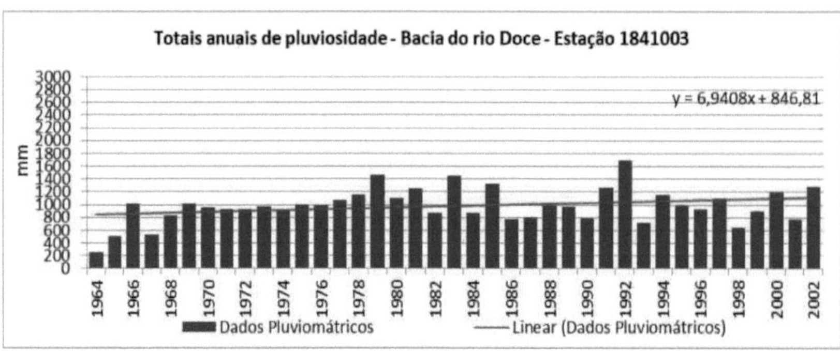

Figure 12 Annual rainfall totals at station 1841003

Station 1842008 - Santa Maria do Suaçui

Located in the state of Minas Gerais, in the southwestern region of the DOCE 01 sub-basin, it has data from 1971 to 2002. With rainfall variations ranging from 650.7 to 1782.8 mm per year, it has a tendency towards rainfall stability, with the straight line varying from 1100 to 1200 mm on average.

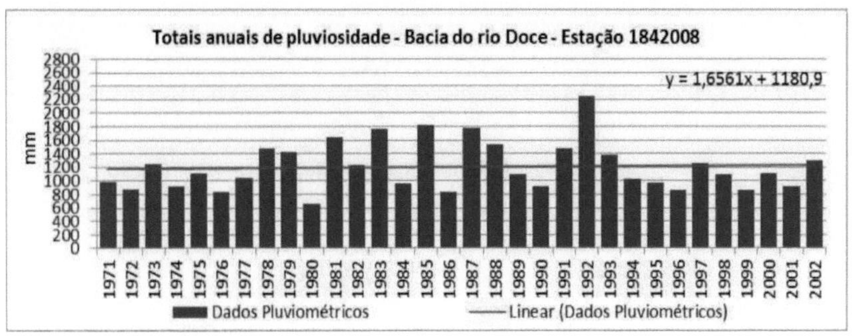

Figure 13 Annual rainfall totals at station 1842008

1940000 Station - Itarana

Located in the state of Minas Gerais, in the southwestern region of the DOCE 01 sub-basin, it has data from 1948 to 2002. With rainfall variations from 676.7 in 1963 to 1728.9mm in 1979. The linear average is around 1100 mm.

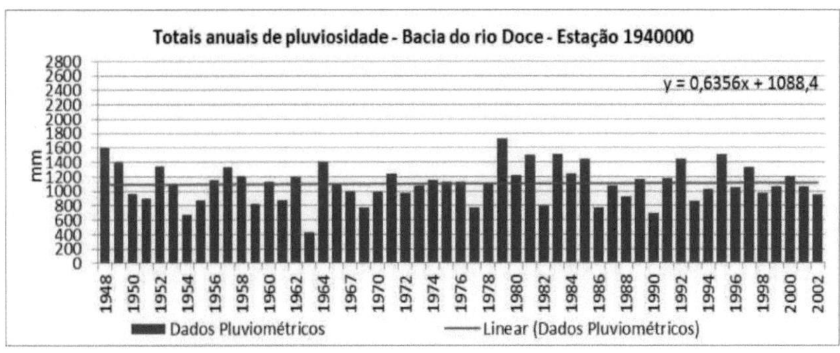

Figure 14 Annual rainfall totals at station 1940000

Station 1940009 - Pancas

Located in the state of Minas Gerais, in the northeastern region of the DOCE 01 sub-basin, it has a clear trend of stability by linear regression. The lowest annual sum is present in 1963 with 352.4mm and the maximum in 1981 with 1753.9mm.

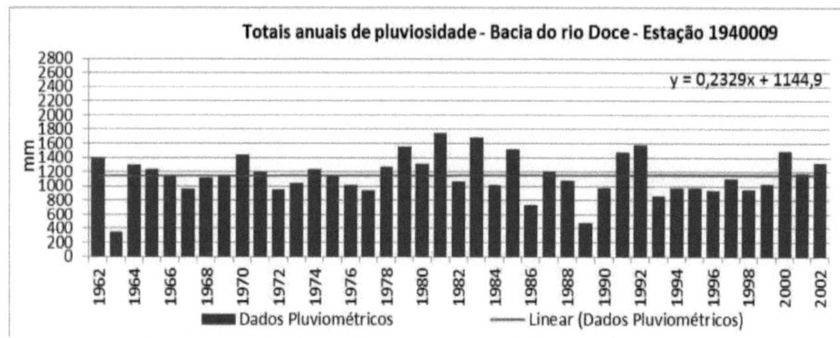

Figure 15 Annual rainfall totals at station 1940009

Station 1940012 - Itaimbé

Located in the state of Minas Gerais in the southern region of the DOCE 05 sub-basin, it has data collected from 1958 to 2002. The highest rainfall total was in 1979 with 1635.1mm and the lowest in 1963 with only 250.8mm.

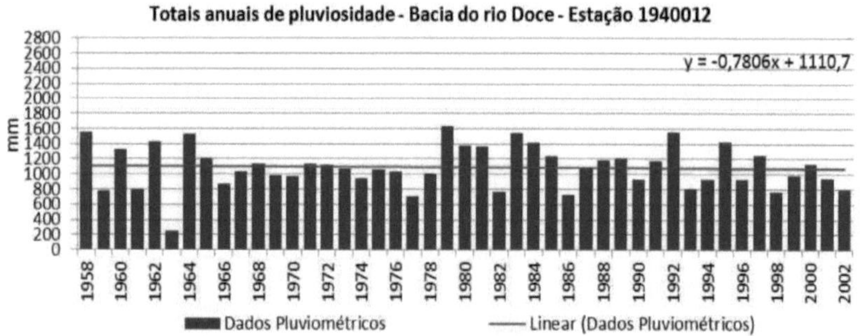

Figure 16 Annual rainfall totals at station 1940012

4.1.3 Group 3 - Downward trend in annual rainfall totals

Station 1840000 - Aguia Branca

Located in the state of Minas Gerais, in the southern region of the DOCE 04 sub-basin, the Aguia Branca station has a total of 33 years of data collected. Its maximum annual rainfall was 1919.2mm in 1983 and its lowest was 690.4mm in 1993. The downward trend is evidenced by the linear variation from 1500mm to 1200mm.

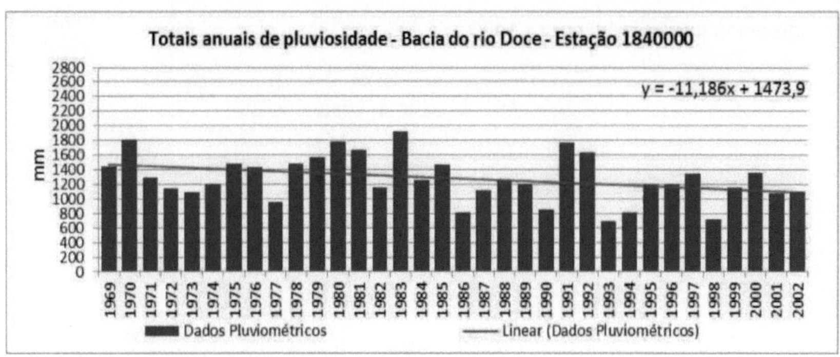

Figure 17 Annual rainfall totals at station 1840000

Station 1841004 - Governador Valadares

The Governador Valadares station is located in the southwestern region of the DOCE 01 sub-basin. It has the lowest and highest annual rainfall in two consecutive years: 1963, as shown by other stations, was the driest year on record in the Rio Doce basin, with 263.7mm and 1964 with 2622.4mm. At this station, the linear regression varies from 1600mm to 1300mm.

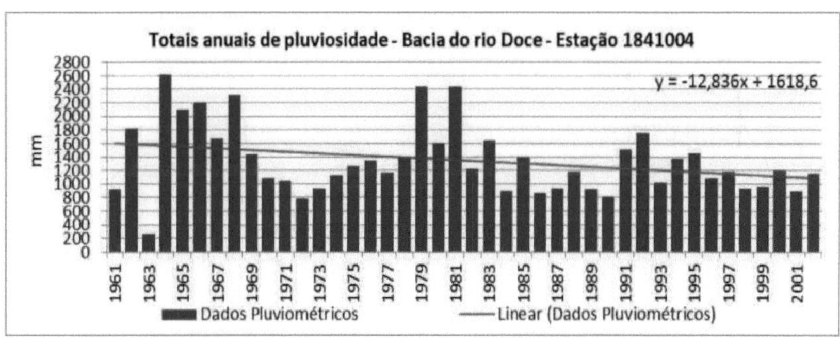

Figure 18 Annual rainfall totals at station 1841004

Station 1842005 - Coroaci

The Coroaci station is located in the southwestern region of the DOCE 01 sub-basin in the state of Minas Gerais. With a total of 32 years of data, the station recorded its highest rainfall in 1979, with a value of 2303.7mm, and its lowest in 1997, with a total of 676mm.

The linear regression varies subtly from 1400 to 1300mm.

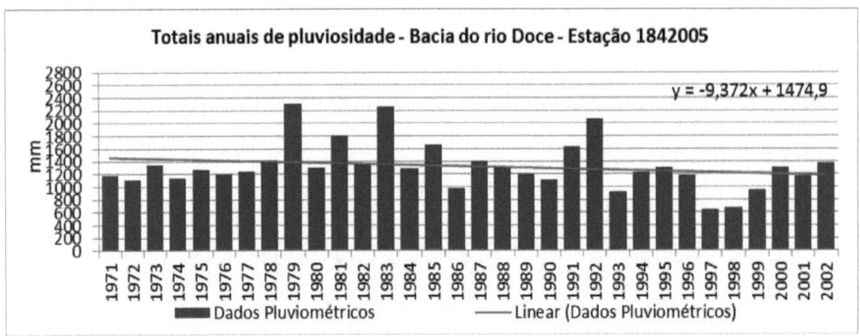

Figure 19 Annual rainfall totals at station 1842005

Station 1842007 - Guanhaes

Located in the western region of the DOCE 01 sub-basin, the Guanhaes station in the state of Minas Gerais confirms a downward trend in rainfall in this sub-basin, specifically in the western and southwestern regions. The linear regression is most evident at this station, decreasing from 1600mm to approximately 1200mm.

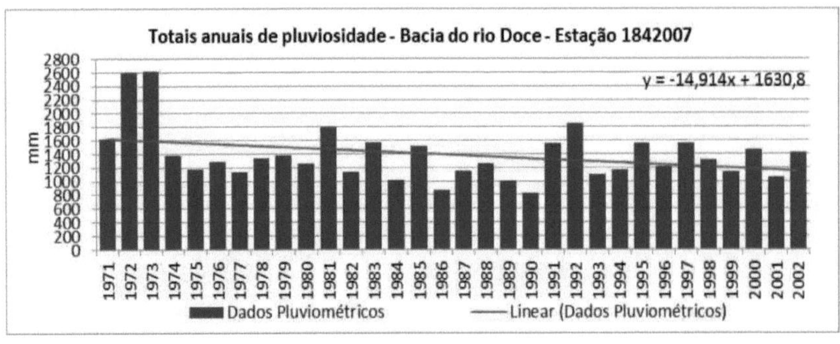

Figure 20 Annual rainfall totals at station 1842007

Station 1940020 - Caldeirao

Located in the state of Espirito Santo, in the southern region of the DOCE 06 sub-basin, the station has a total of 32 years of data. The highest average rainfall was recorded in 1979 with 1770mm and the lowest in 1990 with 721.2mm. The falling linear regression ranges from 1500mm to 800mm.

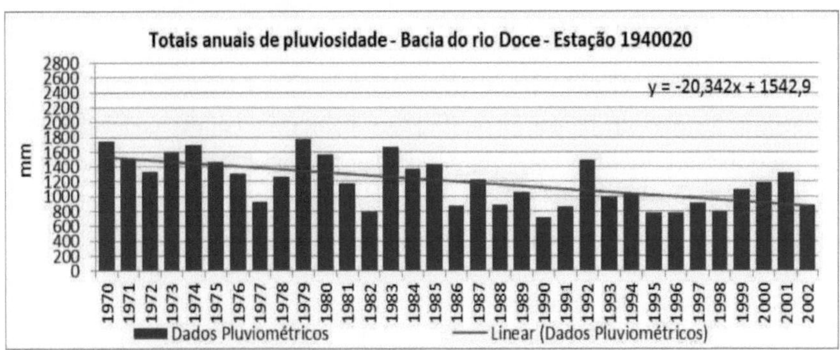

Figure 21 Annual rainfall totals at station 1940020

Station 1943008 - Santa Maria do Itabira

Located in the state of Espirito Santo in the southwestern region of the DOCE 06 sub-basin, the station has collected a total of 60 years of data. It has the highest annual rainfall in 1952 with 2365.3mm and the lowest in 1963 with 794.4mm, once again confirming this year as the driest in the Rio Doce basin in the data collected at the stations for this research. The linear variation is between 1600mm and just under 1200mm.

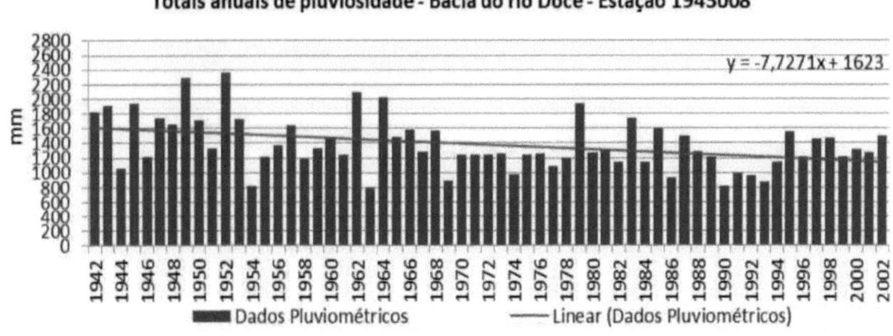

Figure 22 Annual rainfall totals for the 1943-2008 season

4.2 . Analysis of interannual variability

4.2.1. Deviations from annual average values

In this section, the graphs will show the relationship between the amount of rainfall in each year in relation to the average rainfall for all the years in the period of each rainfall station. They therefore indicate whether there were significant negative or positive anomalies. Likewise, given the impossibility of listing all the rainfall stations, some graphs have been selected which are representative of the situation in the basin as a whole.

Station 1841004 - Governador Valadares

From the perspective of linear regression analysis, the Governador Valadares station shows a tendency for the amount of rain to decrease. An almost cyclical behavior can be found in relation to the variability of the series. At the start of the analysis, in the period from 1964 to 1969, the averages are positive and in these years we find the highest annual rainfall averages for the station. This was followed by five consecutive years of negative averages from 1970 to 1975. From 1986 onwards, the negative averages are the majority, justifying the linear regression with a downward trend in annual rainfall totals for the station.

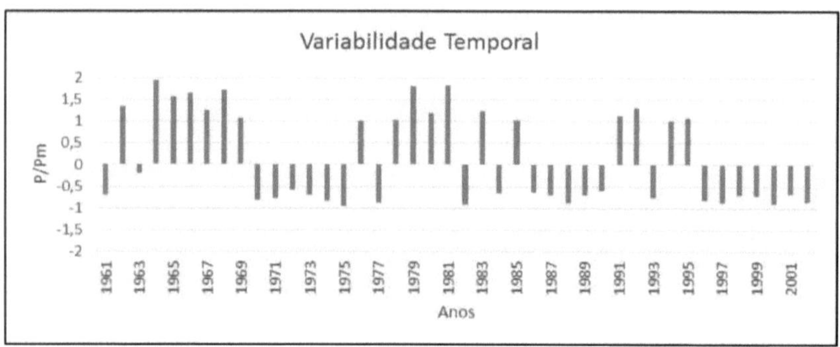

Figure 23 Temporal variability at station 1841004

Station 1842007 - Guanhâes

Located in the southwestern region of the basin, the Guanhâes station shows a tendency for the annual rainfall totals to decrease according to the linear regression analysis. The temporal variability graph shows a total of 20 years of negative averages over a period of

31 years of data.

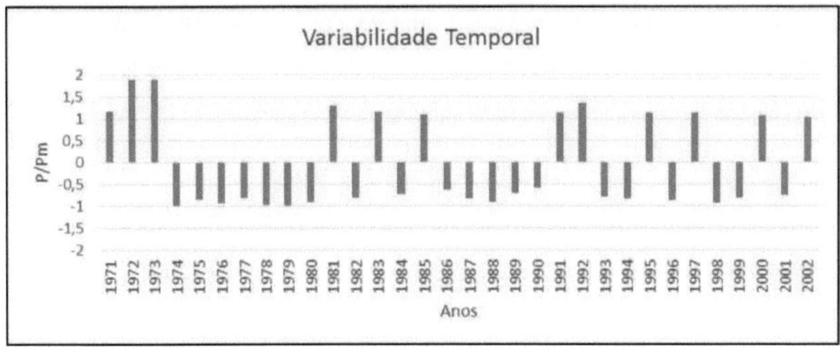

Figure 24 Temporal variability at station 1841004

Station 1842008 - Santa Maria do Suaçui

Also located in the southwest of the basin, station 1842008 has a stable rainfall trend according to linear regression analysis. With regard to temporal variability, a total of 15 stations have positive averages and 17 have negative averages. We can divide the total of 31 years of data into three main periods: from 1971 to 1977, except for 1973, the averages are all negative. From 1978 to 1993 we find a significant period of positive rainfall averages and from 1994 to 2002, seven years have negative averages similar to the beginning of the analysis in the 1970s, indicating the start of a possible rainier period.

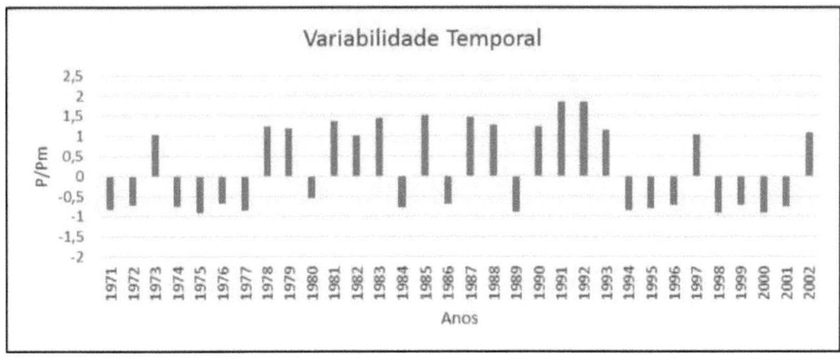

Figure 25 Temporal variability in the 1842008 season

Station 1940009 - Pancas

Located in the southern region of the basin, the Pancas station also shows a trend towards stable rainfall from the perspective of linear regression. Interpreting the temporal variability graph, we find a balance in the number of years with positive and negative averages between 1962 and 1977. From 1978 to 1987 there was a predominance of positive averages. From 1988 to 1999 there is a clear period of negative averages which seems to begin to change from 2000 onwards with three consecutive years of rainfall totals above the seasonal average.

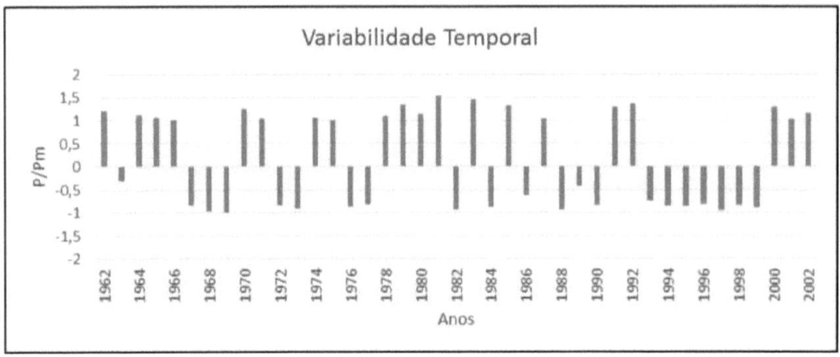

Figure 26 Temporal variability at station 1940009

Station 1940012 - Itaimbé

Located in the south-central region of the basin, the Itaimbé station, according to the linear regression analysis, has a tendency towards stability and a slight drop in annual rainfall. In line with this analysis, the temporal variability graph does not show a clear trend. It can be seen that in the first 20 years, most of them, 13 years in total, show below-average rainfall. In the following 24 years, 13 years have positive averages and 11 have negative averages, confirming the linear regression trend.

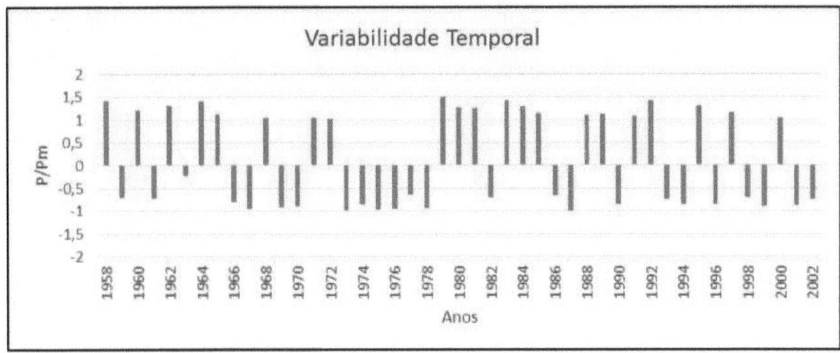

Figure 27 Temporal variability at station 1940012

Station 1940020 - Caldeirao

Located in the southeast of the basin, despite showing a linear trend of decreasing rainfall, from the point of view of temporal variability, we don't find a definite pattern. In the first 15 years of data, 13 were above average rainfall. From 1986 onwards, we have a period of 14 years in which 8 of them consecutively show below-average rainfall. Perhaps this is the reason for the downward trend in rainfall.

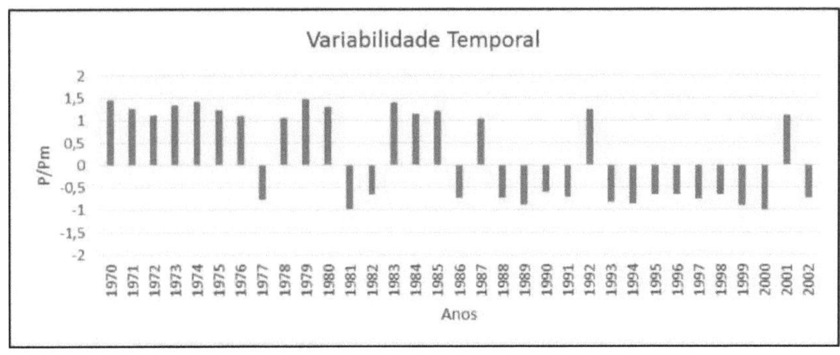

Figure 28 Temporal variability at station 1940020

Station 1942002 - Bom Jesus do Galho

Located in the northwest region of the basin, the station shows a tendency towards a slight increase in rainfall from a linear regression perspective. The temporal variability graph does not show a clear trend, but if we divide it into two periods, from 1942 to 1975 there are more years with below-average rainfall totals. From 1976 onwards, the values are mostly above average. In this way, it is understood that the rainfall regime varies in cycles, or periods, making it arbitrary to project a clear trend.

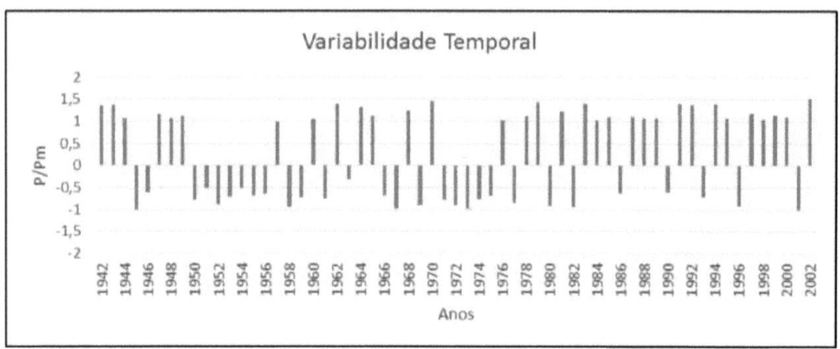

Figure 29 Temporal variability in the 1942002 season

Station 1943001 - Piracicaba River

Located in the eastern region of the basin, in the state of Espirito Santo, the station shows a slight increase in rainfall totals under linear regression analysis. In the temporal variability graph, it can be seen that 29 years have seen rainfall totals above average and 30 below average. Despite this balance between the number of years, from the 1980s onwards there is a clearer trend with most years above average rainfall. This period probably brings a linear upward trend.

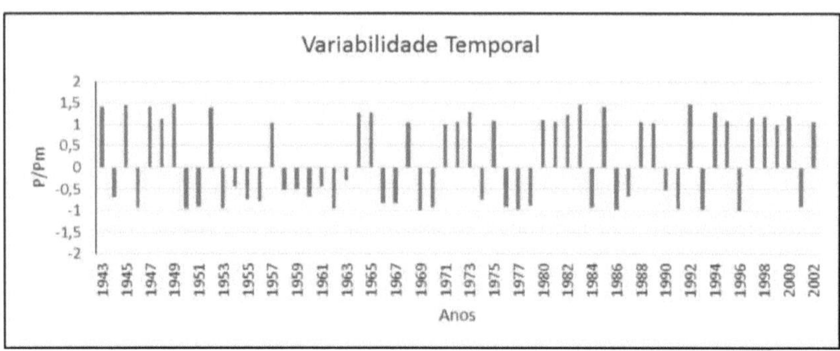

Figure 30 Temporal variability in the 1942002 season

Station 2043010 - Piranga

Located in the state of Espirito Santo, in the eastern region of the basin, the Piranga station has a tendency to increase rainfall from a linear regression perspective. In this graph, it can be seen that in the first 21 years the values are predominantly below average, characterizing a less rainy period in the locality. From 1978 onwards, in 24 years, only 4 years recorded below average rainfall, characterizing a period with more rain. It can therefore be concluded that the rainfall regime is extremely variable and relatively cyclical.

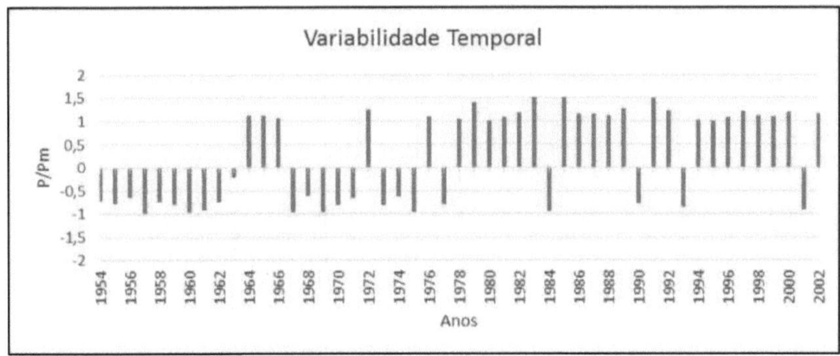

Figure 31 Temporal variability at station 2043010

Over the years, the rainfall stations studied show variability, with more rainy periods and less rainy periods, making it dangerous to state with certainty a specific trend for each station and for the basin as a whole.

4.2.2 Standard deviation analysis

Table 1 lists the stations with their respective codes, followed by the average rainfall total for each station and the average standard deviation, which indicates the variability of the rainfall totals in relation to the total average for each station. Values closer to zero indicate that there was no variability, i.e. that in the season in question, the rainfall amounts were almost the same as the average. Higher values indicate much higher or much lower rainfall than the average. In the following columns, the average plus and minus the standard deviation were calculated. Then, for each station, the number of years below and above the average was counted, characterizing a greater or lesser number of rainy years.

Table 1 - Annual averages, deviations and anomalous years

Station	P	DP	P+DP	P-DP	Positive Years	Negative years
1840000	1360,2	82	1442,2	1278,2	15	18
1841001	1317,5	247,6	1565,1	1069,9	23	27
1841003	617,4	368,2	985,6	249,2	17	21
1841004	1342,6	207,7	1550,3	1134,9	18	24
1842005	1247,3	72,8	1320,1	1174,5	11	20
1842007	1384,7	116,2	1500,9	1268,5	12	20
1842008	1208,2	109	1317,2	1099,2	15	17
1940000	1349,6	244	1593,6	1105,6	26	28
1940001	1177	138,5	1315,5	1038,5	26	19
1940005	1427,2	102,3	1529,5	1324,9	24	29
1940006	1030,3	62,2	1092,5	968,1	14	16
1940009	1149,8	127,9	1277,7	1021,9	20	21
1940012	1092,7	231,7	1324,4	861	21	24
1940013	1067	92,2	1159,2	974,8	13	20
1940016	1124,3	21,7	1146	1102,6	14	20
1940020	1197,1	273,4	1470,5	923,7	16	17
1940023	1276,2	63	1339,2	1213,2	15	17
1940025	1169,7	72,5	1242,2	1097,2	14	18
1941005	1087,3	140,1	1227,4	947,2	20	20
1941006	1042,9	28,2	1071,1	1014,7	25	20
1941008	1114,2	83,5	1197,7	1030,7	28	26
1941009	899,6	4,2	903,8	895,4	13	21
1941010	997,4	39,2	1036,6	958,2	17	17
1941011	1288,8	106,8	1395,6	1182	13	19
1941012	931,9	56,7	988,6	875,2	18	14
1942002	1089,8	200,3	1290,1	889,5	32	29
1942006	1536,3	367	1903,3	1169,3	26	29
1942008	1173	11,8	1184,8	1161,2	14	18
1943001	1373,9	284,7	1658,6	1089,2	29	30
1943002	1309,2	116,8	1426	1192,4	13	19
1943003	1082	143,7	1225,7	938,3	16	16
1943007	1639,2	277,2	1916,4	1362	30	30
1943008	1600,9	217,5	1818,4	1383,4	24	36
1943025	1390,6	198,6	1589,2	1192	29	26
1943027	1553,4	84,6	1638	1468,8	27	28
2041008	1163,8	120,7	1284,5	1043,1	29	26
2041023	1334,7	239,8	1574,5	1094,9	17	17
2042008	1501	291,9	1792,9	1209,1	25	35

2042010	1389,9	217,2	1607,1	1172,7	27	33
2042011	1483,1	104,2	1587,3	1378,9	28	32
2043009	1482,9	104,4	1587,3	1378,5	28	32
2043010	1330,7	258,7	1589,4	1072	26	23
2043011	1560,1	160,2	1720,3	1399,9	29	31
2043014	1392,8	5,7	1398,5	1387,1	29	31
2043025	1301,1	18,2	1319,3	1282,9	19	23
2143003	1841,7	285,4	2127,1	1556,3	31	29

Table 1 Annual averages, deviations and anomalous years

The graph in figure 32 shows the average rainfall for each season. From these values, it was possible to identify the values and spatialize the highest and lowest averages, making it possible to understand the rainfall dynamics within the basin. The lines in the graph represent, in orange, the rainfall plus the average standard deviation values and, in gray, the rainfall values minus the average standard deviation.

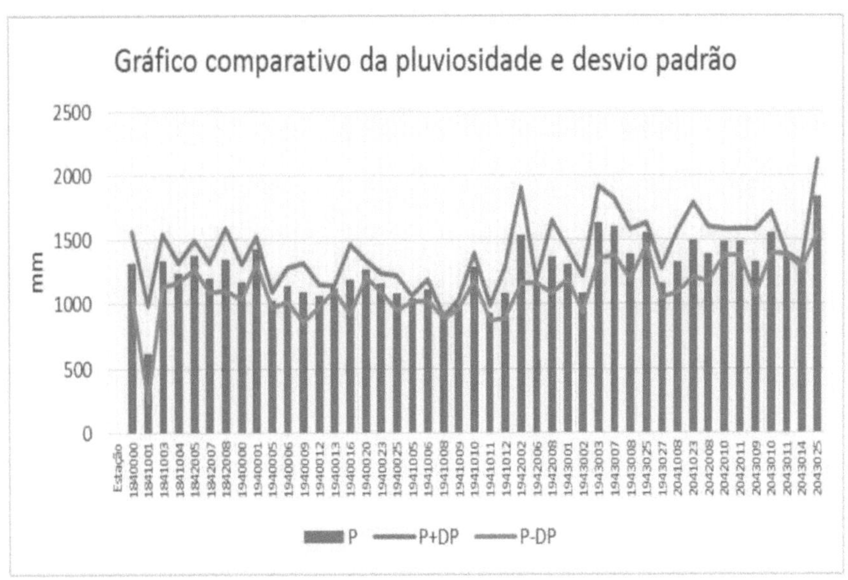

Figure 32 Comparative graph of rainfall and standard deviation

Figure 33 shows the standard deviation graph, which shows two stations with the highest standard deviation, Vila Matias Station 1841001 and Bom Jesus do Galho Station

1942002. At the former, rainfall was found to be below average and at the latter, well above average. Another six stations have a standard deviation above 250: Barra de São Gabriel 1940016, Dom Cavati 1942008, Ferros 1943003, Afonso Clàudio 2041023, Acaiaca 2043009 and Usina da Brecha 2043025. These stations have rainfall averages ranging from 1200 to 1600mm.

There are six stations in the standard deviation series that are close to zero. They are: Novo Brasil 1940013, Barra do Cuieté 1941005, Laranja da Terra 1941008, Vermelho Velho 1942006, Fazenda Paraiso 2043011 and Porto Firme 2043014. Values close to zero indicate less variability, i.e. values closer to the average.

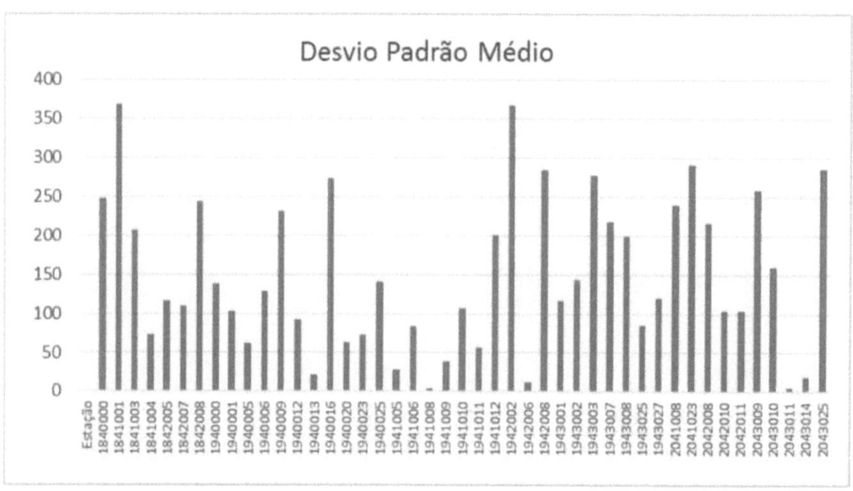

Figure 33 Standard deviation graph

5. CONCLUSION

In order to obtain a meteorological pattern for the Doce River basin using some statistical parameters, it was observed that rainfall variability is quite significant. The average annual rainfall in the basin is 1270 mm over the 61 years studied.

The rainfall behavior of the Doce River basin is influenced by global and regional climate dynamics. Rainfall distribution is not uniform across the study area. The highest rainfall averages are located in the western region of the basin, a fact that can be related to the orographic aspects of the region, where there are high altitudes that lead to the so-called orographic rains. The interaction of the atmospheric systems with the orography in the western sector of the basin intensifies the rainfall in the region.

The basin has a marked rainy season. The rainfall regime of the Doce River basin, from the highest rainfall totals to the lowest, is characterized by a west-east direction, i.e. from the Espinhaço and Mantiqueira mountain regions to the coast. In the central portion towards the coast, there is also a reduction in rainfall totals. This discontinuity is due to atmospheric forces such as the South Atlantic Subtropical Anticyclone and the Northeastern Cavado, which cause a strong process of air subsidence, inhibiting the mechanism that favors rainfall. These forces, interacting with the configuration of the local relief, mean that these areas, located in the central region of the basin, have the lowest average rainfall (Cupolillo, 2008).

With the methodology developed for this research, it was not possible to pinpoint a clear long-term trend in rainfall in the basin. With the historical study of the data, it can be concluded that rainfall variability is intense and, at some rainfall stations, follows a cyclical pattern. Over the years, the stations surveyed show variability, with more rainy periods and others less, making it dangerous to state with certainty a specific trend for each station and for the basin as a whole.

In short, this research has made it possible to verify the spatial and temporal variability of rainfall, presenting an assessment of the situation by analyzing historical data. With the methodology used, it is not possible to identify a clear future trend in the rainfall regime, and there is a need for further research, especially in terms of methodological tools.

6. REFERENCES

AYOADE, J. O. *Introduçao à climatologia para os tropicos*. Rio de Janeiro: Bertrand Brasil, 1996. 332 p.

Doce River Basin Committee. Available at: <http://www.riodoce.cbh.gov.br/PlanoBacia_PIRH-Doce.asp> Accessed on August 6, 2014.

CUPOLILLO, F. *Hydroclimatological diagnosis of the Doce river basin*: spatialization and related dynamic aspects. 2008. Thesis (Master's in Geography) - Federal University of Minas Gerais, Belo Horizonte, 2008.

FERREIRA, V. O. *Trend analysis in rainfall series: Some methodological possibilities.* REVISTA GEONORTE, Special Issue 2, V.1, N.5, p.317 - 324, 2012.

Fonseca, G.A.B. *The vanishing Brazilian Atlantic Forest.* Biol. Conserv. v.34, pp.17 - 34, 1985.

SOS Mata Atlàntica Foundation & INPE (National Institute for Space Research). 2001. *Atlas of Atlantic Forest remnants and associated ecosystems 1995-2000.* SOS Mata Atlàntica Foundation and INPE, Sao Paulo.

GUERRA, C. B.; BARBOSA, F. A. R. *Programa de educaçao ambiental na bacia do rio Piracicaba*: Curso básico de formação de professores na área ambiental na bacia do rio Piracicaba. Belo Horizonte: FNMA/UFMG/ICB, 1996. 251 p.

Brazilian Institute of Geography and Statistics. Available at: <http://www.ibge.gov.br/home/> Accessed on July 10, 2014.

LAGE, M. R.; CUPOLILLO, F.; ABREU, M. L. *Climatic aspects of the Doce river basin.* In: SIMPÒSIO BRASILEIRO DE GEOGRAFIA FÌSICA APLICADA, XI, 2005, Sâo Paulo.

Mittermeier, R.A.; Filho, A.F.C.; Constable, J.D.; Rylands, A.B. & Valle, C. *Conservtion of primates in the Atlantic Forest of Eastern Brazil.* Int. Zoo. Yearbook v.22, pp. 2 - 17, 1982.

MORETTIN, P. A.; TOLOI, C. M. C. *Time Series Analysis.* 2. ed. Sâo Paulo: Edgard Blücher, 2006. 538p.

NIMER, E. *Climatologia do Brasil.* Rio de Janeiro: Fundaçâo Instituto Brasileiro de

Geografia e Estatistica, 1989. 421p.

NIMER, E.; Brandâo, A. M. P. M. *Balanço hidrico e clima na regiao dos Cerrados*. Rio de Janeiro: Fundaçâo Instituto Brasileiro de Geografia e Estatistica, 1989. 421p.

ROSS, J. L. S. & DEL PRETTE, M. E. *Water Resources and Watersheds: Anchors of Environmental Planning and Management.* Journal of the Department of Geography/FFLCH/USP. N° 12, 1998.

SILVA, P. O. M. P.; GOLDSCHMIDT, R. R.; SOARES, J. A.; FERLIN, C. Forecasting Time Series Using Fuzzy Logic. 4th CONTECSI - University of Sao Paulo, 2007.

SILVA DIAS, P. L.; MARENGO, J. A. A. Atmospheric waters. In: TUNDISI J. G.; REBOUÇAS, A. C.; BRAGA, B. *Fresh waters in Brazil*: ecological capital, use and conservation. C.E.M.; Hidrologia: Ciência e Aplicaçao. Sao Paulo: Escrituras Editora, 2002. p.703.

STRAUCH, N. (Org.) *A bacia do rio Doce*: Estudo Geogràfico. Rio de Janeiro: Fundaçao Instituto Brasileiro de Geografia Estatistica, 1955, 199 p.

STRAUCH, N. (Org.); *Zona metalûrgica de Minas Gerais e vale do rio Doce*. Rio de Janeiro: Conselho Nacional de Geografia, 1955, 192 p.

TAVARES, A. C. Climate Change. In: VITTE, A. C.; GUERRA, A. J. T. *Reflexoes Sobre a Geografia Fisica no Brasil*. Rio de Janeiro: Bertrand Brasil, 2004. 280 p.

Printed by Books on Demand GmbH, Norderstedt / Germany